Written and Illustrated by

L. Auerhahn

Lucky Penny Press
Nurturing the Creative Spirit

ISBN 978-1938136610

First Edition
10 9 8 7 6 5 4 3 2 1

Text and Illustrations by L. Auerhahn

Published by Lucky Penny Press, a divison of Lucky Penny Publishing
1776 Park Avenue, Suite 4-369
Park City, UT 84060
info@luckypennypress.com • www.luckypennypress.com

A portion of the proceeds of Li'l Ghost will benefit Summit County Library
www1.youseemore.com/SummitCounty/

Dedication:

For Tysha, LaLa, & Little Lou, witches extraordinaire. Limster, a mischievous monster in his own right, my eccentric Mad Scientist, "The Don", Mario the Magnificent – fixer of all and practical prankster, and of course my Mummy, the most loving, thoughtful, self-less soul, I've ever had the privilege of knowing. May I someday become the person you've all taught me to be – I love you!

Disclaimer:

All pranks performed by a professional poltergeist! Do not attempt at home – unless under the express permission and supervision of a Mummy, Daddy, or other creature over the age of 118. For further information please consult: "Li'l Ghost's Book of Pranks& Potions"

Alliteration:

Alliteration is the occurrence of the same letter or sound at the beginning of adjacent or closely connected words. This book is full of alliterations - can you find them all?

There is an obscure valley tucked away beyond the woods, concealed from prying eyes –
where primordial things lurk both day and night.
Unknown to mortals it's all the rage, it's the place to be –
if you're a monster, spook, witch, or zombie.
All the creepy creatures that make our hair stand on end and our skin crawl –
come to dwell in Hidden Hollow including the most frightening fiend of all….
Li'l Ghost!

Now it is a well known fact that Witches do not associate with the types of beings who live in the gorge known as – Hidden Hollow. So it was natural for everyone to be a bit leery of Li'l Ghost when she first came to stay.

They weren't about to be fooled by her fashionable stripped stockings and infectious smile. They knew that this playful poltergeist was not all she seemed. True she was funny, witty, and friendly, but that pointy black hat atop her head told everyone she met that Li'l Ghost came from a long line of spell-casters and pranksters. Her sort enjoyed causing mayhem and mischief, lurking in corners and playing tricks that would make even the toughest monster shriek.

Standing innocently by her post she appears to be a rather friendly Ghost. After all she was once a very nice little Witch, the youngest in her coven – till she met with a particularly perilous potion that ended in disaster when accidentally dropped into an overheated cauldron.

In time the inhabitants of Hidden Hollow realized Li'l Ghost was simply a good-humored, lighthearted spook, with just the right amount of mischievousness. Before long Li'l Ghost had charmed her way into the hearts and homes of all those who lived in the valley.

Many moons have waxed and waned since Li'l Ghosts first day in Hidden Hollow. She enjoys spending her nights in her favorite stomping grounds – Gregorian's Graveyard – plotting and planning pranks and hexes to play on unsuspecting mortals, surrounded by her favorite friends; Jack, Frank, Marco, Vlad, Will, and Bob.

Li'l Ghost loved that her friends took pleasure in pulling practical jokes on humans. Tricking was always so much more of a treat when done with her comrades in crime. Used to always having her cronies by her side – one can't really blame Li'l Ghost for what was to follow – after a fateful day left to her own devices.

Immune from most ghastly ailments Li'l Ghost became a lone spook when a break out of the Monstrous Mumps sent each and every one of her cohorts straight to their beds. Bored with nothing to do, formulating frightening shenanigans was not much fun when it was just you. Soon her mind began to wonder, and some of her old impish Witch ways began to sneak back through the cracks. In no time at all she was rapidly devising some very troubling tomfoolery to play, not on mere mortals, but on acquaintances and friends!

The ideas were emerging and developing quickly. It seemed that in an instant she'd filled up her, "Book of Pranks & Potions", ready to launch her attack on Hidden Hollows unsuspecting inhabitants.

She grabbed a bolt of invisible wire and ran towards the Wailing Woods. Tied each end to two enormous thousand year old Oaks then hid behind a Deadly Night Shade bush. Sure no one would discover her there, she covered her ears, gave a yowl that would set anyone within a mile on edge, and counted to ten.

"...One, Two, Three, Four..." she began but didn't get to five before a deafening crash caused the earth to tremble and shake. A roar of rage that could be felt vibrating through the heavens soon followed. If you've ever heard the shriek of a Sasquatch you'll comprehend the severity of the situation – if not – well that's what imagination is for.

When the ground and the sky finally settled, Li'l Ghost peeked out from her shelter. Shy Sasquatch Silas was down but not a single tree had fallen. Frustrated her hoax didn't have the domino effect that she had hoped – Li'l Ghost snuck away in search of her next quarry without a thought for poor Si.

Several days later, while passing through Gregorian's Graveyard, on her way to the mausoleum to see Marco, now that the Monstrous Mumps epidemic had passed, Li'l Ghost saw the gravediggers turnip lantern bobbing through the cemetery and happened upon a thought. Unfortunately for Gedwyn, she acted on it.

Flitting from tombstone to tombstone she snuck up behind Gedwyn and blew out the flame. Neither one of them saw the open grave in front of the gravedigger. By the time Gedwyn discovered it Li'l Ghost had vanished into thin air – literally. It was a trick she had learned from a certain gifted friend of hers, Invisible Eddie.

She arrived at Marco's and was excited to hear that he could play. She had special plans for just the two of them that night. For years she had wondered what the Mummies looked like under all their wrappings. Were they horribly disfigured? Or had they been burned? Why was Marco always so anxious about his bandages coming undone? There seemed to be a relentless battle between him and his textile coverings. They were constantly unraveling, and Marco was continually rewrapping, tucking in ends here and there.

While Marco was distracted by a cluster of fire flies Li'l Ghost grabbed hold of a stray piece of cloth and tied it to a sturdy tree. Unknown to Marco the iridescent insects were there to do her bidding and he quickly found himself in a lightning bug battle.

Sidling up to the distraught boy she whispered, "If you move your arms up and down while spinning super fast – like a windmill – they'll fly away."

"Really? Gee – thanks Li'l Ghost – you're such a great friend!" he smiled as he began to twirl faster and faster until…all of the fire flies had disappeared and so had all of his wrappings! Looking towards the numerous strips of cloth now lying in a heap Marco stood there in nothing but his boxer briefs.

Trick
or
Treat

Disappointed to discover that there was 'nothing' there, Li'l Ghost shrugged and said, "I guess that answers that question" and took off running before Marco's Mummy and Daddy could say a word. It was too bad that Li'l Ghost couldn't see his disheartened face, or the tears that welled up in his eyes at being let down by his so-called friend. Maybe if she had, the mayhem that was soon to follow could have been averted, but as it was, the 'pranks' were in motion and heads – or at least one in particular – were about to roll.

How she pulled off her next trick was a wonder to all. The tale gets taller with each telling and has become a legend among the creatures of Hidden Hollow. Truth is no one really knows what happened that fateful night and Li'l Ghost's lips are sealed. However it happened the reality is this; at around midnight on a Thursday evening Li'l Ghost lured the reclusive headless Hessian, Hubert, and his black stallion, Ichabod, out of hiding, where upon she swiftly stole his Pumpkin head!

How she dodged all Hubert's famous fireballs is part of the mystery, even years latter it doesn't take a keen observer to make out the scorched trail of his hot pursuit, now nicknamed, 'Li'l Ghosts Treacherous Trail'. The hollow had never seen chaos like what followed before or since. Of course Li'l Ghost was nowhere to be found. Her success with the Headless Horseman went straight to her head.

Li'l Ghost was unstoppable!

"Cock-A-Doodle
Doooo!"

She switched out the Zombies favorite meal for a vegetarian tofu delight. She figured it had the same texture and they were too dead to notice. In fact she was somewhat pleased with herself as it seemed they enjoyed their new health food craze. Healthy food – healthy brain – right?

Inspired she discreetly substituted the Vampires challises of… well you know… blood – yuck – with Cherry flavored Italian Soda. It was not a success. Let's just say that being chased from the Count's Castle by a colony of blood thirsty bats – not fun! Definitely not on the 'to do again' list.

Discarding any more thoughts of playing with others' food she made her way to the hills where the Woolfurred family was practicing for their annual Howl-A-Thon. Once a year they performed for the hollow to raise money so they could send their son, Will, to the Grand Ole Opry down south. Li'l Ghost leaned up against a tree and listened for awhile before she began her incantation. Her witchy spell transformed the wolves' melodious "Arrrrr-Ohhhhs" into a cacophonous "Cock-a-doodle-dooo"!

FRANK N. STEIN
ABNORM
HUMAN ANATOMY
PATHOLOGY
FIRST AID
CALCULUS
BIOCHEMISTRY
GRAYS ANATOMY
NEURO SCIENCE
VIROLOGY
PHYSIOLOGY
Pocket ATLAS MED
SLEEP DISORDER
OXFORD
MEDICAL

Next on her list was her good friend Frank N. Stein. She knew he liked to get up as early as twilight to study his books on anatomy, medicine, physiology, psychology, and a lot of other 'ologies', Li'l Ghost couldn't pronounce. Frank was working on his medical degree. He wanted to be a family practitioner. His brain was so full of knowledge that it kept 'revolverating' all day long. He couldn't get any sleep, consequently in the late hours of the morning he would put his 'mind to rest' so he could get some shut eye. It was the perfect time for some Li'l Ghost mischief.

Li'l Ghost had acquired an abnormal brain from Frank's partner, Igor, she told him she was doing 'research' on brains for the Zombies. While Frank slept she exchanged the two then stayed up all day merely to watch Frank bumble awkwardly through the night. Li'l Ghost laughed at herself – this was the best trick yet – Frank wasn't even aware he had lost his mind!

The coup de grâce took place on All Hallows Eve. Every goblin, zombie, monster, and ghost, along with other eerie and unearthly things, met at the Widows Well for a toast of Good Luck, before descending upon the mortal world to wreak havoc for one night.

No one suspected anything odd at first but slowly they realized that they were all disappearing! Of course for the spooks, specters, phantoms, and ghouls, this was really no big deal. So they stood by and shared a good laugh with Li'l Ghost as they watched the pandemonium ensue.

It took only an hour or two for the potion she placed in the well to wear off but that was long enough to have caused confusion and disorder disturbing their Halloween plans. The beings of Hidden Hollow had had enough! Before stepping into the mortal realm they all agreed that something drastic had to be done –

Li'l Ghost must be stopped!

Widow's
WELL

A covert council was held, discussions developed, and everyone in attendance was given a mission to manage. They were going to make good use of, 'The Golden Rule', by doing to Li'l Ghost what she had done to them. Everyone went to work immediately. They had a lesson to teach this pesky poltergeist!

Si silently strung string from tree to tree then discreetly watched as Li'l Ghost went flying head over heels, landing face first in her favorite fetid frog pond, while playing, "Catch the Hag Tag".

Vlad and his family concocted an unpleasant potion to replace her much loved Cherry Italian Soda. The bloodless beverage was prepared with ingredients too nasty to mention here. Believe me when I tell you that no amount of brushing teeth or eating monster mints could take out the terrible taste from Li'l Ghost's mouth. Suspended in saliva – it lingered for weeks – Gross!

The Woolfurred's worked with Winifred Witch who helped them cast a small spell. For one week every time Li'l Ghost tried to speak a disgusting green gob floated out – popping with a loud, and very stinky, "Cock-a-doodle-dooo"!

Frank and Hubert worked as a team. Winifred helped them with a creative confounding charm as well – except theirs was only a 24 hour spell. However all things considered that was long enough. For a night and a day Li'l Ghost believed she had lost her head! She ran around confused and befuddled. She didn't know up from down and became lost in Gregorian's Graveyard, walking in circles for hours, looking for her missing noggin!

Distraught as she was her friends did not give up. They had one final lesson. The coup de grâce! The worst trick of all – the silent treatment! To an outgoing, friendly fiend, like Li'l Ghost, being ignored, invisible, and virtually friendless was too much for the small spook. Li'l Ghost ran to her tombstone, sobbing – she had had enough!

Marco watched from atop the mausoleum. His Mummy and Daddy had taught him well. He was a kind, patient, and very forgiving boy and had decided not to participate in the tricks. Marco knew that some night Li'l Ghost would understand that pulling pranks was not polite. It appeared that tonight was that night. Marco recognized that what Li'l Ghost needed the most right now was a friend and that was what he was.

He went to Li'l Ghost, sat beside her, till she looked up, and then gave her a hug. The two talked for hours and Marco agreed to help Li'l Ghost carry out her new plan. For Li'l Ghost had come to realize she had acted awfully rude to her friends and that she needed to make amends. Full of excitement she couldn't sleep so she began that very day.

While Si slept, snoring, not so softly, Li'l Ghost left him a note of apology along with a gift she knew he would treasure. Sweet sapling starts for rare redwood trees!

FIRE
EINSTEIN
TESLA
GALILEO
da VINCI
STEIN
NEWTON
ARISTOTLE
KEPLER
EDISON
FARADAY
SAFETY GOGGLES
#3001
X-TRA plugs, CORDS & Switches
CAUTION
SKELETON PARTS

For Vlad her plan took a little longer. After collecting the rarest blood from around the planet she was prepared to make her presentation. The Vampires, in their bat form, were rendered flightless with surprise.
Gedwyn, the gravedigger, was shocked to discover enchanted floating pumpkins that lit up as he made his rounds and went out as he left. He'd never have to hassle with carrying and lighting, or relighting, his turnip torch again. He thought they looked a lot like Li'l Ghost's pal, Jack, so he called them – Jack O'Lanterns.

One can only imagine the joy and delight on William and his family's faces upon receiving an invitation to audition for the World premiere of Wolf-Gang's Opera, "The Howling Hounds of Hurstville".

Frank and Hubert of course were flabbergasted by their visits from the greatest heads and minds of all time. They spent countless hours picking each other's brains, exchanging ideas, thoughts, and dreams.

As for the Zombies...well everyone agreed that leaving their change in appetite was for the best. After all they were content and blissfully clueless that they'd been tricked. Li'l Ghost did give them a recipe book, "Vegetarian Veggie Feasts", so they could switch things up a bit.

Marco had given Li'l Ghost the best gift ever; help to discover her true self and sincere friendship. A debt she knew she could never repay – but Li'l Ghost was going to try anyway – starting with a brand new crate full of imported muslin wrappings with Velcro ends!

As not to forget the creatures involved in the Samhain mayhem, Li'l Ghost chose a few fiends each night and performed secret acts of service. She left encouraging notes, gave caring words to passersby, or stealthily left a handmade gift at their door.

POTIONS
All HALLOWS EVE
PYROTECHNICS 101
HEXES, SPELLS & CHARMS
EVERY DAY CURSES
HISTORY OF HIDDEN HOLLOW
You're INVITED
31. OCTOBER
HAUNTING, MISCHIEF, MAYHEM,
MIDNIGHT
RSVP

While stirring up trouble had been fun at the time, Li'l Ghost realized that it was much more enjoyable to give small acts of kindness and service to her fellow fiends. In fact she was surprised to discover that there was a great deal more she could do. Li'l Ghost's mind was so full of unique ideas that she could barely contain her anticipation at putting them into action!

Li'l Ghost found pleasure and joy in being benevolent, compassionate, and kind, and was ashamed that she had ever thought that pulling pranks was cool. So imagine her confusion when the desire to play tricks tickled the back of her mind once more. She knew jokes could be hurtful and had made a promise to herself and her friends to never be rude or mean to them again – what was Li'l Ghost to do?

She spoke with Jack, Marco, Frank N. Stein, and the rest of the gang, including Bob, (although he didn't contribute much). How could she keep her promise? After all she had once been a tricky witch, before becoming a prank playing poltergeist. Tomfoolery was in her blood! She knew that she'd eventually succumb to mayhem and mischief and that something had to be done.

Not to worry – her cronies came to her rescue by reminding Li'l Ghost how she used to spend her nights before allowing her mischievousness to go astray. Now Li'l Ghost is content with spending her days dreaming of doing good deeds for her monstrous mates, and using her nights to scheme and plot with her comrades in crime. They plan all sorts of spells, hoaxes, and tricks to play on our unsuspecting souls' sphere. (In other words plotting to play tricks on the unassuming world of mortals – that's you and me.)

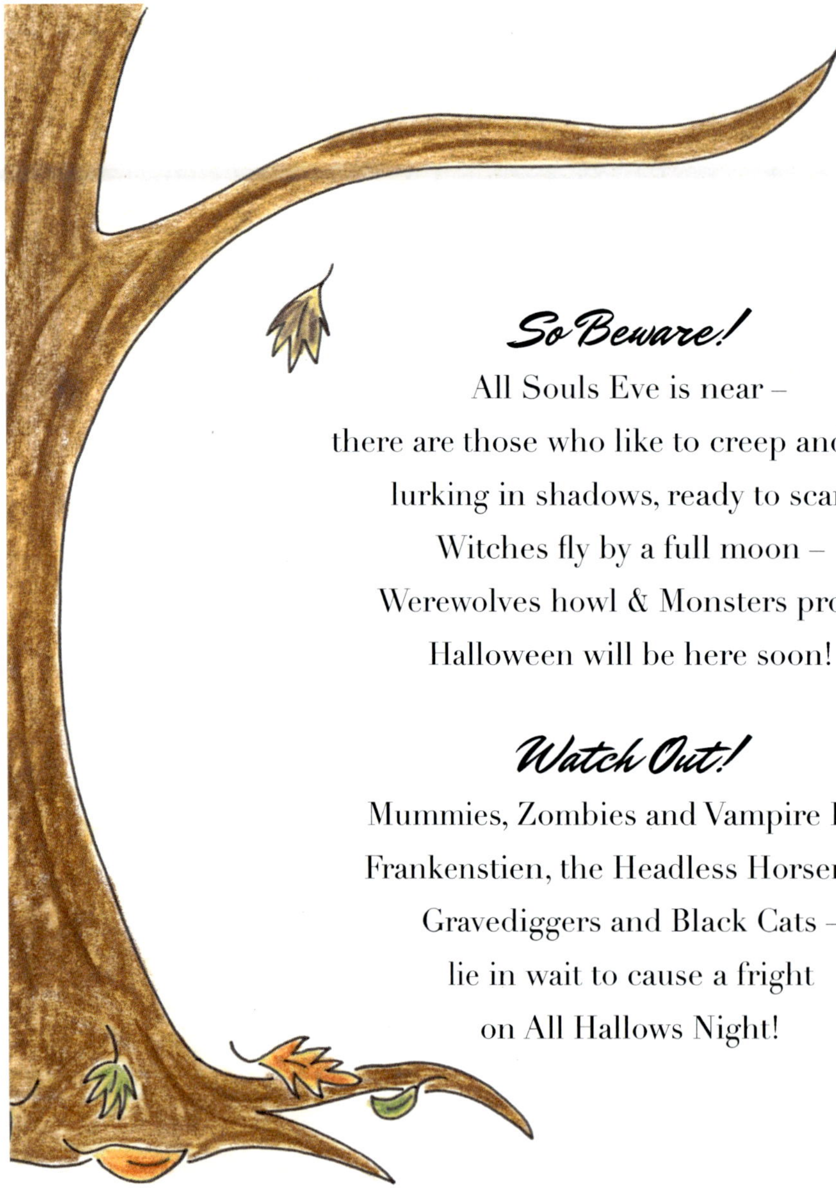

So Beware!

All Souls Eve is near –
there are those who like to creep and crawl,
lurking in shadows, ready to scare!
Witches fly by a full moon –
Werewolves howl & Monsters prowl,
Halloween will be here soon!

Watch Out!

Mummies, Zombies and Vampire Bats,
Frankenstien, the Headless Horseman,
Gravediggers and Black Cats –
lie in wait to cause a fright
on All Hallows Night!

Take Heed!

For Samhein is here,
bonfires are lit, masks are on,
the games and intrigues have begun –
goblins, spooks, and brutish beasts are near,
the worst of the worst is heading your way,
though this specter is rarely seen –
she's stealing into our realm just to say,
Happy Halloween!

Who is this poltergeist all fiends fear the most?
Why it's none other than Hidden Hollows –
Li'l Ghost!

Happy
Halloween

Fun Facts – Did You Know:

Samhein pronounced SAH-win, is a Gaelic Festival marking the end of the harvest season and the beginning of winter. It is celebrated from sunset on the 31st of October to sunset on the 1st of November. Historically Samhein was celebrated in Ireland and Scotland; it was the time when cattle were brought down from their summer pastures. Bonfires were lit, and offerings of food and drink were made to nature spirits to help ensure that the people and their livestock would survive winter. Mumming and Guising (dressing up in masks and costumes), were part of the festival and involved going door-to-door, reciting verses in exchange for food like nuts and apples.

All Hallows Eve was celebrated on the 31st of October in preparation for All Saints' & All Souls' Day.

All Saints' Day is a Christian festival in honor of all the saints, celebrated in the West on the 1st of November.

All Souls' Day is a festival celebrated in some Christian churches with prayers for the Souls of the Dead, especially their ancestors, held on the 2nd of November. Over time, Samhain, All Saints' Day, and All Souls' Day, merged to create modern day Halloween!

A bunch of bats are called a Colony or Cloud of Bats, when crows decide to gather it is called a Murder of Crows, and although most owls like to hang on their own when a crowd of them meet up it is called a Parliament of Owls.

Sasquatch (also known as Bigfoot) is the name given to a large, hairy, bipedal, humanoid creature, said to inhabit forests.

The Hosts of Hidden Hollow

Li'l Ghost was the youngest and most promising witch in her coven. She had an extraordinary knack for pranking and potion making. Preoccupied with cramming for her Witch's Weekly Hex Test, she didn't pay close enough attention to a particularly perilous potion. Too late she realized that the vile of, 'Bella & Donna's Deathly Draft', she was setting in the cupboard, missed the shelf, fell into the cauldron, and exploded. Now she spends her time planning covert operations to hex, charm, and trick all those who dwell in the human realm, with her friends Jack, Frank N. Stein, Marco, Vlad, Will, and sometimes Bob.

Jack was an old brown turnip torch who'd been thrown out with the rubbish. Li'l Ghost felt sad for him so she took him home. She soon discovered that his favorite color was orange and that he used to light the way guiding an old gravedigger from grave to grave. With an effortless enchantment she turned Jack orange, with his permission of course, Li'l Ghost takes him everywhere – the two are inseparable – best friends!

Frank N. Stein is the cleverest creature Li'l Ghost has ever met. Frank's quick, sharp wit and intellectual insights have provided hours of fascinating discussion between the two. Frank is Li'l Ghosts go to person when she needs extra help with anything out of the ordinary. Frank is studying to be a doctor and takes care of spooky ailments remarkably well.

Marco lives with his parents, Menkaura & Muminah, in the mausoleum near the cemetery where Li'l Ghost and Jack like to hang out. They are quiet and keep to themselves. Marco will come and play with them sometimes but his Mummy likes to keep him close by. He doesn't talk much – I guess they have secrets they like to keep under wrap!

Vlad is practicing to take up the family trade and is a bit too ostentatious for Li'l Ghost. However, every now and then his pretentious personality comes in handy. Li'l Ghost does love his accent and thinks he makes a brazen bat, good for eavesdropping and passing covert messages back and forth.

William Woolfurred IV is a ton of fun to play with and is often Li'l Ghost's cohort enjoying a good cloak-and-dagger game. He's like an energetic puppy, always jumping and moving, ready to play with the pack. However he gets a bit moody around the time of the full moon. He wants to be a famous opera star. He has a great voice and boy can he howl!

Silas (who prefers his nick name, Si – pronounced with a long I) is a very shy Sasquatch and keeps mostly to himself. He lives in the woods and is very protective over the flora and fauna of the forest. His best friends are; the Felix Fox family, sister hedgehogs – Helga, Hattie, & Hazel, and a parliament of Owls, one in particular – Lir – who likes to ride on Si's shoulder.

Bob sort of just shows up every now and then, he seems to be tired a lot, and has a very limited vocabulary. His lack of personality is too boring and lifeless for an active ghoul like Li'l Ghost. Bob tends to meander without any real purpose and is easily distracted by people with brains.

Hubert was a soldier in the German military until he met his demise in a secret skirmish. Hubert was very handsome, a favorite of all the girls. Beloved by his mother and his father's preferred son. His brothers in arms honored him for his legendary skills with a blade both on and off the battle field. Unfortunately it was his blade that got the best of him while practicing a top-secret move during an encounter with an unnamed fiend. Being decapitated by your self is not only a difficult task but it is rather embarrassing for a decorated swordsman. One can only imagine the humiliation poor Hubert feels at the accidental loss of his head.

Winifred is a tricky witch who likes helping others with hexes and spells, especially if it causes chaos, confusion, and overall havoc. Not much is known about her whereabouts or coven which is true with all the witches of Hidden Hollow as they pretty much keep to themselves.

L. Auerhahn can relate more with Frank than with Li'l Ghost. She loves to learn, read, and write, and is not much of a prankster. Although she's had her share of being the brunt end of a trick or two as her husband is a practical joker. Like Li'l Ghost she enjoys giving and receiving small acts of kindness especially in secret. She too lives in a small hollow surrounded by li'l monsters and critters that go bump in the night!

THE END

Thank You!

First and foremost – L. Auerhahn would like to thank her Heavenly and Earthly families for her heritage and talents. Auerhahn family your unconditional love and support has been a wonderful blessing. Thank you to her Pearce family who give without hesitation to strangers and family alike. Their willingness to give the shirts off their backs – quite literally – has been a blessing in desperate times and their leftovers have filled more than just stomachs on numerous occasions. Being the recipient of many secret acts of service has made L. Auerhahn a very grateful ghoul indeed. Thank you to her Ridgeland Ward Family where her family of Li'l Spooks grew up with more love than they knew what to do with. They've never found a better gaggle of people and they never will! She'd like to express her gratitude for Pam and Megan, who are the best of the best, and who were the first to smile and extend a hand of friendship to a family of 'move-ins' in a not too welcoming new town. L. Auerhahn doesn't have the words (amazing as it seems) to express her gratitude for a great friend, Amanda, an amazing author who will no doubt be famous some day. A simple, "Thank you for sharing, laughing, crying, and cheering this author on", just doesn't seem to cut it. She'd like to say "Thank You!" to Melissa Marsted and Lucky Penny Press for taking a chance on her. Thank you to Pamela Beverly-Quigley for designing the book. As always L. Auerhahn gives special thanks to her Mummy & Daddy without whom she would not be where she is today.

Li'l Ghost is the first book in the Hidden Hollow series of stories ranging from Toddler Board Books, and Children's Picture Books, to Beginning Chapter Books and YA Novels. Look for more stories from Hidden Hollow coming your way soon!

About Lucky Penny Press

Lucky Penny Press, a division of Lucky Penny Publishing, is a children's eBook and print publishing company. Lucky Penny publishes books written by both adults and children with some of our books published in multiple languages including Spanish, French, and Chinese and with audio recordings.

All of the books advocate messages and lessons that focus on nurturing the creative spirit, introducing new cultures, and empowering children to believe in their dreams. Many of the themes of our books focus on stories of adventure, nature, culture, and the environment.

As part of Lucky Penny Press's cultural fabric, each book is connected to a non-profit organization that receives a portion of the proceeds.

Lucky Penny Press
Nurturing the Creative Spirit

1776 Park Avenue, Suite 4-369 | Park City, Utah 84060-5148 | info@luckypennypress.com

25278129R00024

Made in the USA
Middletown, DE
24 October 2015